CAN YOU HEAR A SHOUT IN SPACE?

Questions and Answers About Space Exploration

BY MELVIN AND GILDA BERGER

ILLUSTRATED BY VINCENT DI FATE

SCHOLASTIC REFERENCE

CONTENTS

KEY TO ABBREVIATIONS

cm = centimeter / centimetre
kg = kilogram
km = kilometer / kilometre
km^2 = square kilometer / kilometre
l = liter
m = meter / metre
t = tonne
°C = degrees Celsius

Text copyright © 2000 by Melvin Berger and Gilda Berger
Illustrations copyright © 2000 by Vincent Di Fate.
All rights reserved. Published by Scholastic Inc.
SCHOLASTIC and associated logos are trademarks and/or registered trademarks of
Scholastic Inc.

No part of this publication may be reproduced, or stored in a retrieval system, or transmitted in
any form or by any means, electronic, mechanical, photocopying, recording, or otherwise,
without written permission of the publisher. For information regarding permission, write to
Scholastic Inc., Attention: Permissions Department, 555 Broadway, New York, NY 10012.

ISBN 0-439-09583-2 (pb)

Book design by David Saylor and Nancy Sabato

10 9 8 7 6 5 4 3 2   0/0 01 02 03

Printed in the U.S.A.   08
First printing, February 2000

Expert Reader: Thomas D. Jones, Ph.D.,
NASA, Houston, Texas

For the Erly brothers—Brian, Steven, and Andrew
— M. AND G. BERGER

For the late John Wood Campbell, Jr.,
a legendary figure in science fiction literature
— V. Di FATE

INTRODUCTION

Most of us will never go into space. Yet space exploration plays a big part in our lives. We read stories about space travel. We watch movies about space adventures. We see shuttle launchings and landings on television. We play space games—and even have favorite space toys.

Even more, we live better because of space exploration. At the touch of a button, we get live television from far-off places. In seconds, we make satellite phone calls to anywhere in the world. And if we sail boats or fly planes, we use satellites to find our position and to help us navigate.

Everything we hear and see about space makes us curious to know more. How do rockets blast shuttles into space? What is it like to walk on the moon? How do astronauts eat, sleep, and go to the bathroom in space? Why do astronauts wear space suits? What is the *International Space Station*? Have space probes found life on any planet? Who chooses and trains astronauts?

Can You Hear a Shout in Space? answers all these questions—and many, many more. Read on, if you're curious—or if you dream of a trip to space one day!

Melvin Berger *Gilda Berger*

STEPS INTO SPACE

Can you hear a shout in space?

Not unless you have a radio in your space suit. There is hardly any air in space. Without air, or another way to carry the sound waves, there is no sound.

Where does space begin?

Beyond Earth's atmosphere—about 60 miles (96 km) above the surface of the earth. The atmosphere is like a blanket of air. It covers the whole earth and provides the oxygen we need to stay alive. The higher you go, the thinner the air gets. When you reach 60 miles (96 km) there are few traces of air. This is about where the atmosphere ends—and space begins.

Where does space end?

It doesn't. Space is endless. It goes on way beyond the solar system and past the most distant stars.

Can airplanes fly into space?

No. Airplanes need air to fly. Air moving around the wings makes a lifting force that keeps the plane up in the air. And oxygen in the air lets the fuel burn in the plane's engines. Without air, a plane would fall and crash to Earth. That's no way to get into space, you'll agree!

Mercury

Venus

Earth

Moon

How can you get into space?

By rocket. Only a rocket can provide enough power and go fast enough to overcome Earth's gravity and enter space.

A rocket doesn't need wings to fly. It uses its own thrust to climb. A rocket also carries its own supply of oxygen. This means it can burn fuel far out in space where there is hardly any air.

What makes a rocket go?

The burning fuel. It creates hot gases under great pressure. The gases have nowhere to go—except out through a small opening at the bottom of the rocket. The gases rush out very fast and with great force. The rushing hot gas shoots the rocket up into the air. A rocket can move much faster than the speed of sound!

How can you make your own rocket?

Blow up a balloon and hold the neck tightly closed. Then let go. Watch the balloon shoot this way and that as the air inside rushes out the opening.

The air in the balloon is like the gases in the rocket. As the air rushes out, it sends the balloon flying in the opposite direction. Unfortunately, your balloon-rocket won't get too far!

Who invented the first rockets?

The Chinese, about 800 years ago. Their rockets were powered by gunpowder. People set the gunpowder on fire. The gunpowder burned quickly or exploded and sent out a burst of hot gas. The burning gas shot the rocket into the air.

At first, the Chinese used rockets for fireworks. Later, they attached arrows to the rockets and used them in war. In the 1200s, soldiers in the Chinese army fired these early rockets at their enemies.

Which was the first modern rocket?

The small rocket launched by Robert Goddard in 1926. The rocket shot up only about 40 feet (12 m) in 2 seconds and then fell down. Yet the launch made space history. It was the first rocket to use liquid fuel.

What was the first long-range rocket?

The V-2, first launched in 1942. During World War II, the Germans used the V-2 to carry a large bomb to a distant enemy target.

After the war ended in 1945, United States scientists learned to build their own rockets. The new rockets went as high as 250 miles (400 km) into space.

When did the Space Age begin?

October 4, 1957. On that day, the Soviet Union launched a rocket that carried *Sputnik* (later called *Sputnik I*) into space. *Sputnik* was the first satellite made by humans to circle our planet.

Sputnik flew around Earth in a loop called an orbit. The satellite took just over an hour and a half to complete its orbit. The size of a soccer ball, *Sputnik* weighed 184 pounds (83 kg). It carried a radio transmitter that kept beeping. The beeps helped scientists on Earth track *Sputnik*'s position in space.

What keeps a satellite in orbit?

Its great speed. Satellites in low orbits travel at an incredible 17,500 miles (27,350 km) an hour! But at the same time they are moving forward, gravity is pulling them down. The combined motion—forward and downward—keeps the satellite in orbit. It remains at a safe altitude above Earth—and doesn't even need engines to stay up!

Sputnik

What was the first living being in space?

A small dog named Laika. On November 3, 1957, Laika was lifted into orbit in *Sputnik II*. She completed seven orbits. Then her oxygen supply ran out and she died. Still, Laika made history. She proved that it is possible for a living being to survive in space.

Who was the first person in space?

Yuri Gagarin of the Soviet Union. He blasted off on April 12, 1961, and completed one orbit of the earth at an altitude of $203\frac{1}{5}$ miles (327 km). Gagarin spent close to two hours in space.

The American astronaut Alan Shepard Jr. was next, on May 5, 1961. But Shepard didn't go into orbit. His spacecraft rose to an altitude of 115 miles (184 km). It returned to Earth 15 minutes and 28 seconds later—the shortest spaceflight in history.

Shuttle orbiter
with fuel tanks

Apollo/Saturn 5 moon ship

Who was the first American to orbit Earth?

The astronaut John Glenn Jr. A rocket sent his spaceship, *Friendship* 7, into space on February 20, 1962. Glenn completed three orbits at an altitude of 162 miles (260 km) in just under five hours. You can see *Friendship* 7 at the National Air and Space Museum in Washington, D.C.

In October 1998, at age 77, Glenn went up into space again. At the time, he was the oldest person ever to do so.

Who was the first woman in space?

Valentina Tereshkova of the Soviet Union. She started her flight on June 6, 1963. Over the next three days, Tereshkova made 49 orbits of the earth.

Mercury capsule with escape tower

V-2 rocket with mobile launcher

Lunar lander

11

Do astronauts feel gravity inside their spacecraft?

No. But it's not because they are beyond the pull of gravity. It's because of the way their craft is flying forward and falling at the same time that the astronauts feel weightless in orbit.

You may feel the same way for an instant inside an elevator that suddenly starts going down very fast. Because you and the elevator are falling at the same speed, you get the feeling of being weightless. If you were standing on a scale in the elevator, the scale would register zero at the moment the elevator dropped.

Who took the first "space walk"?

Aleksei Leonov of the Soviet Union, on March 18, 1965. Leonov spent 10 minutes floating outside the *Voskhod 2* spacecraft.

Who was the first person to walk on the moon?

The American astronaut Neil Armstrong, on July 20, 1969. On stepping out onto the moon, Armstrong spoke these famous words: "That's one small step for man, one giant leap for mankind."

What was it like to walk on the moon?

Not too difficult. The moon is much smaller than Earth, and its gravity is only one-sixth as strong. Armstrong found that he could jump easily on the moon, even though he wore a heavy space suit. In fact, hopping was the best way to get around.

Suppose you weigh about 60 pounds (27.2 kg) on Earth. Your weight on the moon would be only 10 pounds (4.5 kg)!

Why was landing on the moon so important?

It was the first time that a human being had visited another object in space. Thousands of men and women working together had scored one of the greatest triumphs of all time!

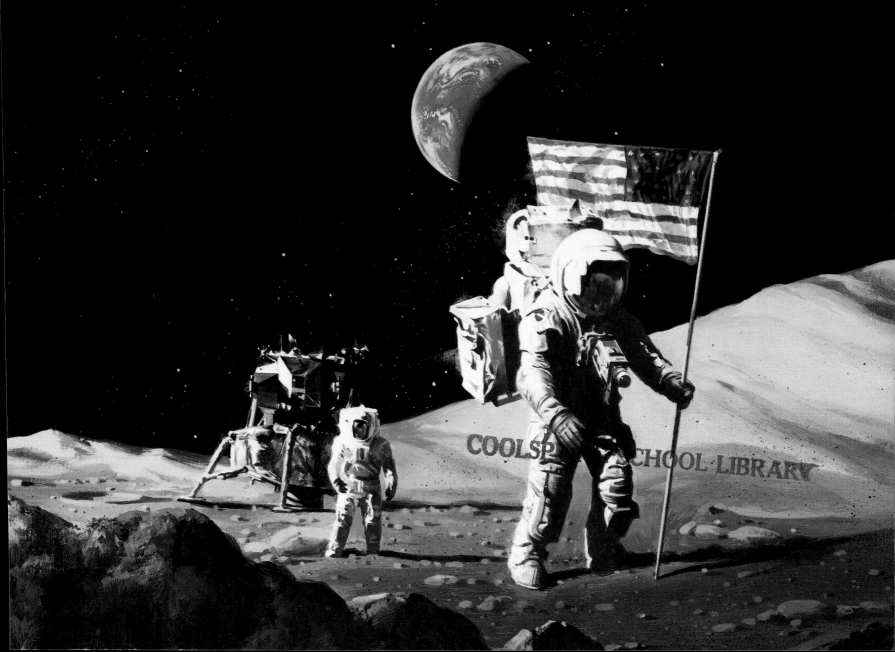

Have there been other moon landings?

Yes. There were six moon landings between 1969 and 1972. Altogether, 12 astronauts have landed on the moon. On the last three landings, astronauts explored the moon with a buggy called the lunar rover.

What did the astronauts bring back from the moon?

Hundreds of pounds of moon rocks and soil. The rocks are brown and gray, with bits of glass in them. Scientists believe many came from ancient volcanoes, since there are no active volcanoes on the moon today.

The rocks also show no sign of plant or animal life. Measurements of rocks from Earth and from the moon suggest that they both formed about 4.6 billion years ago.

What did astronauts leave on the moon?

Instruments to measure the climate and conditions on the moon. The results show that it is boiling hot when the sun shines and freezing cold when it's dark. There also seems to be some shaking caused by weak "moonquakes."

The astronauts also left an American flag at each landing site. Since there is no wind on the moon, they put stiff wire into the flag to hold it open.

Of course, the astronauts left their footprints behind. Since there is no water, air, or life on the moon to disturb them, the footprints will stay there for millions of years.

Besides all that, the mission left behind six lunar landers, three rover vehicles, and lots of leftover tools and equipment. Worth millions of dollars, it's some of the most expensive junk in history.

How many human beings have gone into space?

Altogether, almost 400 men and women. In the United States, the National Aeronautics and Space Administration (NASA) selects and trains all the American astronauts.

How does NASA get astronauts ready for space?

Intensive training. The astronauts-to-be take classroom courses in many subjects, from aeronautics to the effects of space on the human body. They also get flight training in a jet plane, which develops skill in making quick decisions and working as a team member in a complex machine.

Along with flight training, the trainees experience weightlessness. They float weightless in the padded cabins of large aircraft doing sharp climbs and free-fall dives. They also train underwater in a huge swimming pool where they feel almost weightless as they float in the water.

Astronaut training also teaches them to survive an emergency landing. The men and women who want to go into space learn what to do if they touch down unexpectedly in the ocean or in a far-flung jungle or desert. The training can take more than a year. If trainees perform well, they are finally ready. Sooner or later, most astronauts are assigned a mission in space.

Can anyone become an astronaut?

Yes, but it's not easy. The first astronauts were expert airplane pilots. Now scientists with many different backgrounds are also sent into space.

To become an astronaut, you must be very well educated, in excellent physical and mental health, and able to train long and hard. NASA accepts only the most extraordinary people into the space program.

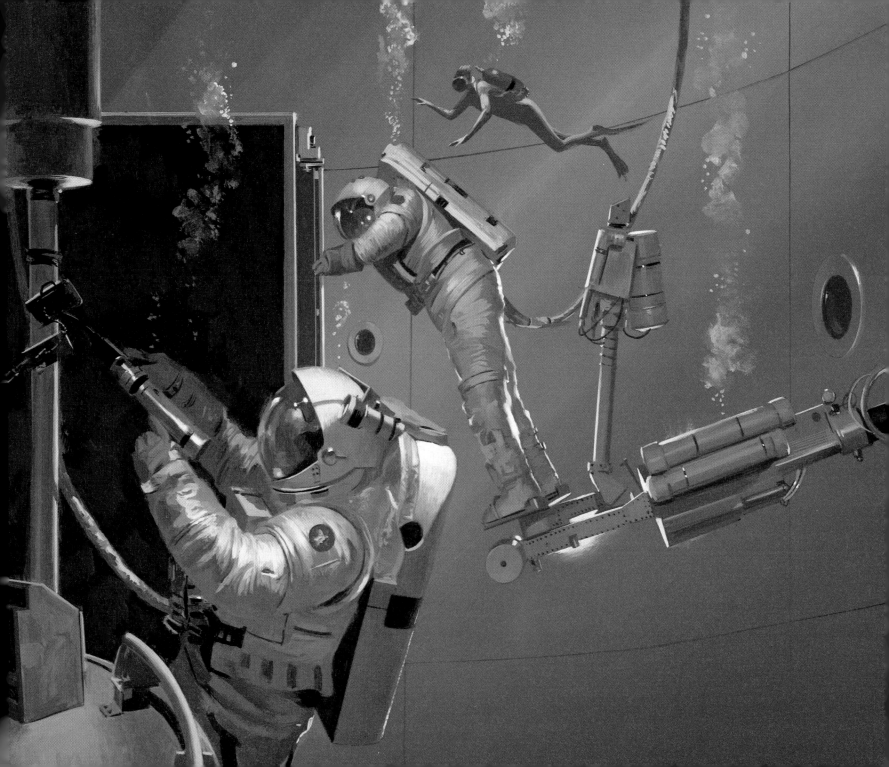

LIVING IN SPACE

When did the "new" Space Age begin?

On April 12, 1981, with the launching of the space shuttle *Columbia*. As NASA put it, the space shuttle is "launched like a rocket, hauls cargo like a truck, and lands like an airplane." The shuttle was the first spacecraft that could fly over and over again. Before that, a spacecraft could be used only once.

Columbia carried a crew of two on its 54-hour mission and completed 36 orbits of Earth. To show that it could fly again, the shuttle went back into space about seven months later.

What does the shuttle look like?

An airplane with attached rockets. The shuttle has three main parts: orbiter, external fuel tank, and two rocket boosters.

The orbiter carries the crew and cargo. As many as eight astronauts can live and work in the cabin. They wear special suits during liftoff and landing, but casual clothes the rest of the time. Behind the cabin is the cargo bay. About the size of a big tractor trailer, it can hold 30 tons (30.5 t) of cargo, called payload. Three main rocket engines provide the orbiter more than 2 million pounds (.9 million kg) of thrust.

A huge external fuel tank, the size of a 15-story building, is attached to the belly of the orbiter. The tank carries fuel for the orbiter's engines.

Two rocket boosters are on the sides of the orbiter. They provide nearly 6 million pounds (3 million kg) of extra thrust to lift the orbiter into space.

External fuel tank

Rocket booster

Rocket booster

Orbiter

USA

NASA

Who flies in a shuttle?

The crew is made up of a commander in charge of the flight, a pilot who helps to fly the orbiter, and several "mission specialists," people who have different jobs, from doing scientific experiments to launching satellites from the shuttle's cargo bay.

How does the shuttle get into space?

Flames thunder out of the three giant engines and two rocket boosters. The force jolts the shuttle off the pad. In two minutes the shuttle is 32 miles (51 km) up in the air. The two rocket boosters cut loose from the orbiter. They drift down by parachute into the Atlantic Ocean, where they are picked up by waiting ships.

After about eight and one-half minutes, the shuttle is at an altitude of 70 miles (113 km). All the fuel in the external tank has been burned up in the orbiter's engines. The fuel tank separates and burns up as it falls down through Earth's atmosphere. This is the only part of the shuttle that is not used again.

What do you feel at blastoff?

The thrust of the rocket engines squishes you against your seat. As the spacecraft speeds up, faster and faster, you feel a force on your body about three times your normal weight. One astronaut said he felt like a gorilla was sitting on his chest!

How long does it take to reach orbit?

Less than 10 minutes. The usual orbit is close to 200 miles (320 km) above Earth.

Does space look blue like the sky?

No. Space is black. Without air to scatter the sun's light, there is no color. From space you can see the planets and stars. But everything around them is pitch-black.

Does it feel good to be weightless?

Not at first. Your sense of balance is upset and you may feel light-headed and giddy. Turn your head quickly, and you feel like you are spinning or tumbling around.

Also, your blood shifts towards your head, your face puffs out, and your nose gets stuffy. Astronauts say that in space you have the same full feeling in your head that you get from hanging upside down from gym bars.

Of course, most people get used to being weightless in a day or so. Or else, how could anyone work in space?

What is it like to work when you're weightless?

Every job takes about twice as long as on Earth. To turn a screw, for example, you have to use arm or foot grips to keep yourself steady. If not, you'll twist around instead of the screw. Using a hammer is just as tricky. A very hard strike can throw you backward. Small wonder astronauts are pooped by the end of the day.

Can you walk in the orbiter?

Not really. To get from place to place you pull yourself along with hand grips. You can also push off from one part of the spacecraft and float to the next part.

Staying in place is even more difficult. You have to hold on to something, put your feet in foot straps, or belt yourself into a chair.

Forget bending over. It's like trying to grab your toes underwater at the deep end of a swimming pool!

Do you get taller or shorter in space?

Up to 2 inches (5 cm) taller. Without the pull of gravity, your spine stretches out. At the same time, your waist shrinks because many of your organs float upward to your upper chest.

Do astronauts exercise in space?

Yes, they do. Astronauts must work out because being weightless weakens their bones and muscles. The exercise takes a lot of time and energy but is necessary. Riding a bicycle in place, for example, forces the muscles to push and pull, just as they do against gravity on Earth. By cycling for 90 minutes, which is one orbit, an astronaut can pedal around the world!

How do you eat, sleep, and go to the bathroom in space?

You eat food from plastic or foil packages. If it weren't kept in a package, the food would float away. Thirsty? Just squeeze juice or water from a foil drink bag into your mouth. If a liquid were in an open glass, it would crawl up the sides and drops would float around the cabin.

You sleep in a sleeping bag that is attached to a wall. If you don't zip yourself in, you'll end up drifting around the cabin.

You go to the bathroom as you do at home—except that you must use handles to hold yourself down on the seat. Air, not water, sucks the waste away.

Can you take a shower in space?

Not on the shuttle; there's no room! But on a space station there is a shower with a special nozzle that squirts water on your body. Then a vacuum-cleaner-like attachment sucks off the soapy water. Just be sure to save enough water to rinse off the soap!

Most astronauts keep clean with sponge baths. They put water and soap on a washcloth and wash that way. They use special soap that does not need rinsing. Drying off with a clean towel is all it takes.

How do you comb your hair in space?

With care. Long hair gets snagged and tangled when you're weightless. So most astronauts wear their hair short—and just give it a quick brush in the morning.

What are the orbiter's main jobs?

To launch satellites. Astronauts in the orbiter also check out satellites before and after they are sent into space. If necessary, they can repair satellites in orbit or return them to Earth for overhaul.

Another job is to build and supply the space station. The orbiter is big enough to haul large pieces of the new station to orbit. The shuttle orbiter also brings new astronauts and fresh supplies to the station.

What is Spacelab?

A complete laboratory that used to fly in the cargo bay of the orbiter. Spacelab was connected to the cabin by a little tunnel. It was a part of many shuttle missions beginning in 1983 and usually spent about 10 days in space.

Scientists conducted many experiments in Spacelab to get ready for the space station's laboratories. They found out how humans, small animals, and plants change while in space. They discovered how different materials behave in weightlessness. A telescope was attached to Spacelab to observe bodies in space without the interference of Earth's atmosphere.

How does the orbiter return to Earth?

The pilot fires two small rockets on the orbiter. The rockets slow the orbiter and nudge it out of orbit. The spacecraft starts to head down toward Earth.

As the orbiter speeds through Earth's atmosphere, squeezed air piling up in front raises the temperature to a scorching 2,300 degrees Fahrenheit (1,260°C). Its belly glows red-hot from the heat. Special tiles on the bottom of the orbiter keep it from burning up.

Closer to Earth, the thicker air works to slow down the craft. Wings allow it to glide to the spaceport. In minutes, it touches down on a long runway. A parachute on the tail opens up to bring the orbiter to a stop.

Satellite

View from the cargo bay of the
orbiter as a satellite is launched.

It all depends. Communication satellites relay television, radio, and telephone signals around Earth. Weather satellites keep watch on conditions in Earth's atmosphere and help make weather forecasts. Mapping satellites study changes on the earth's surface. Secret military satellites track the movement of troops, ships, planes, and weapons. And astronomy satellites study stars, galaxies, and other bodies in space.

Which is the best-known satellite?

An astronomy satellite called the Hubble Space Telescope. Launched in April 1990, the Hubble orbits the earth at an altitude of 380 miles (610 km). The telescope was sent into orbit above the earth's atmosphere so scientists could see farther and have a clearer view out into space.

What is the most famous repair job in space?

Fixing the Hubble Space Telescope. It didn't work perfectly at first. One of its mirrors was a bit too flat. This blurred the images. Also, there was a slight wobble as the satellite traveled in orbit.

In 1993, NASA sent astronauts up in a shuttle to repair the Hubble. They caught the telescope with a 50-foot (15.2 m) robot arm and pulled it into the shuttle's open cargo bay. Working in space suits, they replaced some parts, added new instruments, and launched it back into orbit. Four years later, NASA scientists improved the Hubble even more by attaching several advanced pieces of equipment to the telescope.

Hubble Space Telescope

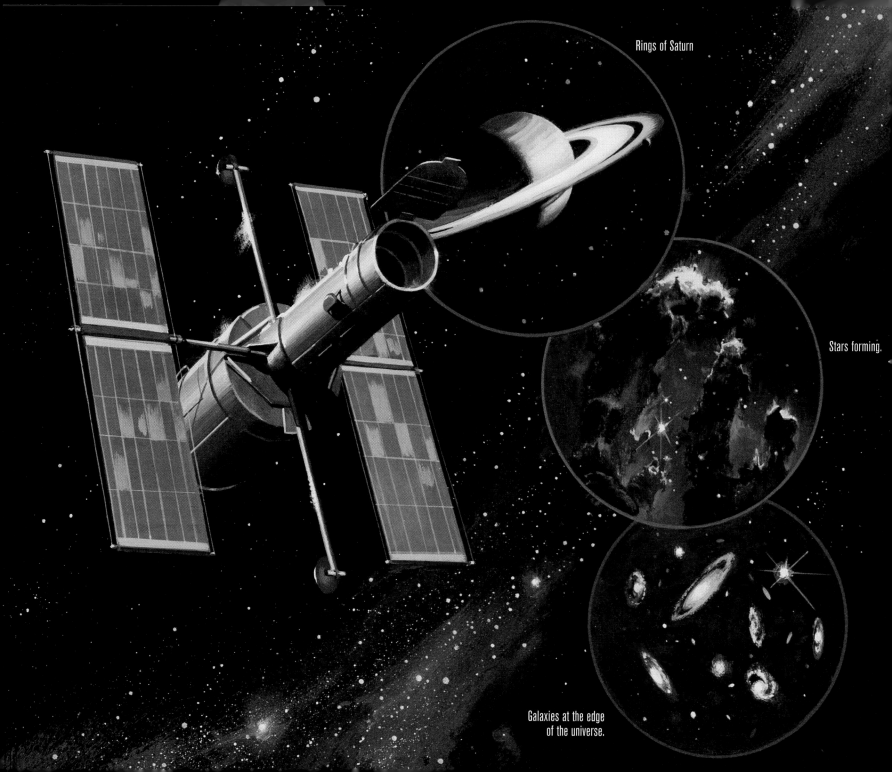

Rings of Saturn

Stars forming.

Galaxies at the edge
of the universe.

Is the Hubble working well now?

It sure is. The Hubble can see seven times farther than the best Earth telescope. It is so powerful that it can see if a penny is heads up or tails up from a distance of 500 miles (800 km)!

The Hubble took the first clear photographs of the planet Pluto. It found clouds of dust and gas that might be forming into new planets around young stars. Over time, it has collected an astounding amount of information about objects billions of miles away in space.

Where do the Hubble and other satellites get their power?

From the sun. Each satellite has solar panels that contain great numbers of solar cells. The cells change sunlight into electricity. Batteries store the electricity and use it to power the satellites when they are in the dark.

How much electricity do the satellites need?

Very little. A typical satellite needs only about as much electricity as an electric toaster!

How many working satellites are in space now?

About 200. Thousands of satellites have been launched over the years. But the rest have either stopped working or fallen out of orbit.

Scientists hope to soon have satellites in space that are smaller and lighter than today's satellites. They will need less power, be more reliable, and perform more functions. One communications satellite, for example, might do jobs as different as spotting insect plagues and tracking air pollution.

What is a space station?

A large satellite that stays in orbit around Earth for a long period of time. Scientists can live and work there for months at a time. Most space stations are found in orbits between 200 and 300 miles (300 and 480 km) above the earth.

What do the astronauts do in a space station?

Experiments, mostly. Astronauts check their own physical condition, develop new materials, test new ways of making medicines, grow different plants and animals, and observe the earth and other bodies in space for weeks or months.

Crew members also spend time taking care of the space station. They clean the living quarters, set up new experiments, prepare for additions to the station, and repair things that break. In their free time, they read, make radio calls home, exercise, watch movies, and take photographs.

What was the first United States space station?

Skylab, sent into space on May 14, 1973. The main purpose was to see how well people could live and work in weightlessness for as long as three months. The crew also conducted many important studies of Earth and the sun.

Who had the idea to study spiders in *Skylab*?

A high school student in Massachusetts. The student suggested an experiment to find out whether spiders could spin webs in weightlessness.

Two spiders, named Anita and Arabella, were carried into space. At first they had trouble making webs. But in a few days, they were back in business building strong and perfect webs. It seems spiders are like humans. They get used to being weightless, too!

Apollo approaching *Skylab*.

Plan for the
International Space Station.

Each one is about the size of a school bus. Space shuttles carry up the modules one at a time. Then astronauts use a robot arm to move them into place and put on space suits to complete the assembly.

Why do astronauts wear space suits outside the shuttle?

Because there's no air in space. Astronauts carry oxygen with them. Tiny fans in an attached backpack provide a steady flow of air for breathing. Automatic valves keep the air pressure inside the suit at a comfortable level. Even so, it's not always easy to move around or do things in the suit. Each one weighs about 300 pounds (136 kg)!

Which is the longest-lasting space station?

Mir, put into orbit by the Soviet Union on February 20, 1986. For 13 years, groups of astronauts from different countries took turns working and living in *Mir*. Valery Polyakov set a record for the longest stay in a space station—one year and 73 days.

Are there plans for other space stations?

Yes. Engineers from 16 countries are building the huge *International Space Station*. A Russian rocket put the first module into orbit on November 20, 1998; the second section went up on the shuttle and was attached a few weeks later.

Future launches will bring up a dozen more modules. When finished, the *International Space Station* will be larger than a football field. The living quarters and labs of the seven-astronaut crew will be as big as the inside of two jumbo 747 jet planes. Some say it is the most complex building project in 100 years.

Which probes went to the most distant planets?

Pioneer 10 and *11* and *Voyager 1* and 2 in the 1970s. All flew past both Jupiter and Saturn and sent back stunning photographs and new information. The probes found volcanoes on one of Jupiter's moons, geysers on one of Neptune's, and strange ice and rock formations on other moons.

Voyager 2 was perhaps most astounding. It flew by Jupiter in 1979, Saturn in 1981, Uranus in 1986, and Neptune in 1989—a full 12 years after its launch in 1977! The probe returned the most detailed pictures of all the planets. It also discovered 10 moons circling Uranus and 6 moons in orbit around Neptune.

Voyager sweeps over
Jupiter's moon, Io.

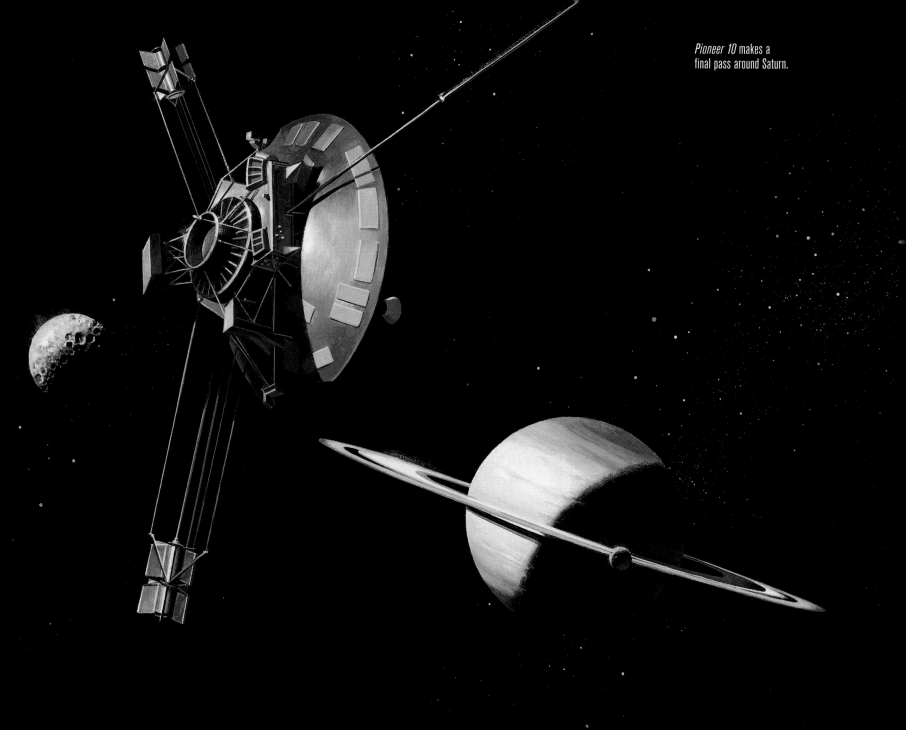

Where are the *Pioneer* and *Voyager* probes today?

Heading out of our solar system. Scientists still hear from *Pioneer 10* and the *Voyagers*. Their travels may continue for millions of years.

Each *Pioneer* probe carries a metal plaque with information about our planet. The plaque shows pictures of a man, woman, and child and also a map of the solar system showing the position of planet Earth. *Voyager* has a recording with music and greetings in various languages. If there are intelligent life-forms in space, we may be hearing from *them*!

Which probe flew close to a comet?

Giotto, hurled into space on July 2, 1985. *Giotto* got within 375 miles (600 km) of Halley's comet on March 13, 1986. A comet is a dirty snowball made up of frozen gases and particles of dust.

At the time *Giotto* was photographing the comet, the probe was traveling 149,133 miles (240,000 km) an hour! The probe is named for Giotto di Bondone, an Italian artist who painted Halley's comet in 1305.

Which probe flew close to an asteroid?

Galileo, which started its six-year journey to Jupiter on October 18, 1989. *Galileo* was heading toward Jupiter when it passed very close to the asteroid Gaspra—a big rock in space. The photos that *Galileo* sent back to Earth gave scientists their first close-up look at an asteroid. The pictures show that Gaspra is only 11 miles (18 km) long.

Do any probes study the sun?

Yes, many do. But the Solar and Heliospheric Observatory, called SOHO, is one of the most important probes in orbit around the sun. The spacecraft, which started its studies in April 1996, measures the atmosphere around the sun and changes in the sun's surface to learn more about what is happening inside that star.

THE FUTURE IN SPACE

Will kids ever travel to space?

Maybe! Every year, space flight is becoming cheaper, safer, and more frequent. In time, a trip into space may be like an airplane flight in the early days of aviation.

Where will the first visitors stay in space?

Most likely in space hotels. A large Japanese company, Shimizu, already has plans for a hotel in space.

The first space hotel will be built inside a giant wheel. The wheel will slowly spin around, creating a feeling of gravity. People will be able to walk around just as they do on Earth. The hotel will have 64 rooms. Guests will go on space walks, take sightseeing trips to the moon, and even play weightless sports!

Will people ever live in space cities?

Very likely. Scientists plan to build huge colonies in space, on the moon, or on Mars. The moon may be a stopping-off point for travelers on their way to far distant planets. Thousands of people will live in these communities for long periods of time.

Are there plans for a new piloted spacecraft?

Yes. A space plane called X-33 may take the place of the space shuttle. It will launch straight up like the shuttle, using a new kind of rocket engine. Once in the air, rockets will boost it into orbit. At the end of its mission, the X-33 will land like a regular plane.

Possible
space hotel.

X-33

Where would people live on Mars?

Inside sealed domes. Factories would create oxygen for breathing from either ice in the soil or from Martian air. Or people would get oxygen from growing plants. In one experiment on Earth, a scientist lived in an airtight room for 15 days with 30,000 wheat plants. The plants took in the carbon dioxide he breathed out and produced oxygen for him to breathe in.

Giant mirrors could reflect sunlight to keep the dome at a comfortable temperature. In the distant future, people could place plants around the ice caps at the North and South poles of Mars. The plants would absorb sunlight and produce enough heat to melt the ice caps and raise the temperature of the entire planet.

How would people on Mars get water?

They could bring it up from Earth. But, according to one scientist, an ounce of water brought to Mars in a space shuttle would cost $10,000!

A much better source is the water frozen in the soil of Mars. People could heat the soil to remove the water. Or they might change water vapor in the air into liquid water.

Finally, people might use huge mirrors made of shiny plastic to reflect sunlight and melt the frozen ice caps. One idea is to use eight giant mirrors, each about a half mile square (1.3 km²)!

Where would people on Mars get food?

The first settlers probably will bring food, just as astronauts do in space stations. Then they'll start growing their own.

People may grow plants in water indoors because plants can't grow outside in Mars's soil. Our neighbor in space has frigid temperatures, no liquid water, and few of the minerals that plants need. Plants grown in water need less space than plants grown in soil. Without bugs around to eat them, plants on Mars might do very well. The biggest crops will probably be peanuts and soybeans.

Will people work in space factories?

Yes. Certain special medicines, metals, and microchips may be made best in weightless space factories. The first "Made in Space" products already exist. They are tiny balls made from liquid plastic. Scientists use the spheres, each as big as a pinpoint, to measure the superfine holes in certain filters. When made on Earth, each ball is not exactly round. The ones made in space are all perfect.

Will we ever get energy from space?

Yes. Scientists plan to send a gigantic panel of solar cells into orbit. The cells would capture the sun's light and change it into electricity. This energy would then be beamed down to Earth.

   The problem is weight. A solar power station would be very heavy. The panels would weigh about 600 times as much as *Skylab*, the heaviest object ever launched into space. Space shuttles might need about 5,000 flights just to carry up all the parts!

What raw materials can we get from space?

Water and metals. Water is an extremely valuable and necessary resource that can be found on some nearby asteroids. And metals to build space colonies may be mined from asteroids or the moon. After all, it would be cheaper to get these materials from space than to haul them up from Earth.

Can we protect Earth from space?

Yes. Over millions of years, a few very large bodies from outer space have slammed into our planet. Some of these crashes have caused tremendous damage.

Scientists today hope to head off any object that seems to be headed toward Earth. One plan is to send up a powerful bomb and explode it near the object. The blast will be strong enough to knock the object out of orbit so it doesn't collide with Earth.

Will we ever get close to another star?

Not for a long, long time. Today, it would take us nearly 200,000 years to get to Proxima Centauri, the next nearest star to our sun.

Imagine that one day it will be possible to travel nearly as fast as light—186,000 miles (300,000 km) a *second*. Even at that fantastic speed, the trip would still take over four years!

How long will we continue to explore space?

As long as

- people want to know more about the worlds beyond our planet Earth.
- scientists keep finding new methods of space exploration.
- there are people brave enough to go off into unknown regions of space.
- people wonder about space and dream of space travel.

INDEX

About the Authors

Mel and Gilda Berger have been following space exploration from the launch of *Sputnik* (1957) to the landing of the first people on the moon (1969) to the start of the *International Space Station* (1998). "For now, we're happy to read and write about space," they say, "and experience space travel vicariously!"

About the Illustrator

Vincent Di Fate was introduced to the wonders of space when he was four years old and went to his first science fiction movie. He was inspired to be an artist by the beautiful paintings of Chesley Bonestell. Mr. Bonestell studied stars and planets through a telescope before he painted them. Mr. Di Fate says, "The paintings took me to those far worlds and made me understand the power of art."